Issues with Access to Acquisition Information in the Department of Defense

A Series on Considerations for Managing Program Data in the Emerging Acquisition Environment

JEFFREY A. DREZNER, MEGAN MCKERNAN, JERRY M. SOLLINGER,
SYDNE NEWBERRY

Prepared for the Office of the Secretary of Defense
Approved for public release; distribution unlimited

For more information on this publication, visit www.rand.org/t/RR3130

Library of Congress Cataloging-in-Publication Data is available for this publication.

ISBN: 978-1-9774-0358-2

Published by the RAND Corporation, Santa Monica, Calif.

Preface

Acquisition data lay the foundation for decisionmaking, management, insight, and oversight of the Department of Defense's (DoD's) acquisition system. Recent statutory changes to organizational structures, as well as roles, responsibilities, and authorities have introduced new challenges and opportunities that significantly affect the collection, storage, and use of acquisition information. This research identifies and describes some of the issues and challenges related to managing acquisition program information in the emerging acquisition environment.

Through a series of short papers, this study informs some of the detailed decisions DoD must make as it implements recent statutory changes to authorities, responsibilities, and organizational structure. These short papers, presented here as chapters, build on six earlier studies on *Issues with Access to Acquisition Data and Information in the Department of Defense*.[1] This report should be of interest to government acquisition professionals, oversight organizations, and, especially, the analytic community.

This research was sponsored by the Office of the Secretary of Defense and conducted within the Acquisition and Technology Policy Center of the RAND National Defense Research Institute, a federally funded research and development center sponsored by the Office of the Secretary of Defense, the Joint Staff, the Unified Combatant Commands, the Navy, the Marine Corps, the defense agencies, and the defense Intelligence Community.

---

[1] Jessie Riposo, Megan McKernan, Jeffrey A. Drezner, Geoffrey McGovern, Daniel Tremblay, Jason Kumar, and Jerry M. Sollinger, *Issues with Access to Acquisition Data and Information in the Department of Defense: Policy and Practice*, Santa Monica, Calif.: RAND Corporation, RR-880-OSD, 2015; Megan McKernan, Jessie Riposo, Jeffrey A. Drezner, Geoffrey McGovern, Douglas Shontz, and Clifford A. Grammich, *Issues with Access to Acquisition Data and Information in the Department of Defense: A Closer Look at the Origins and Implementation of Controlled Unclassified Information Labels and Security Policy*, Santa Monica, Calif.: RAND Corporation, RR-1476-OSD, 2016; Megan McKernan, Nancy Young Moore, Kathryn Connor, Mary E. Chenoweth, Jeffrey A. Drezner, James Dryden, Clifford A. Grammich, Judith D. Mele, Walter T. Nelson, Rebeca Orrie, Douglas Shontz, and Anita Szafran, *Issues with Access to Acquisition Data and Information in the Department of Defense: Doing Data Right in Weapon System Acquisition*, Santa Monica, Calif.: RAND Corporation, RR-1534-OSD, 2017; Megan McKernan, Jessie Riposo, Geoffrey McGovern, Douglas Shontz, and Badreddine Ahtchi, *Issues with Access to Acquisition Data and Information in the Department of Defense: Considerations for Implementing the Controlled Unclassified Information Reform Program*, Santa Monica, Calif.: RAND Corporation, RR-2221-OSD, 2018; Jeffrey A. Drezner, Megan McKernan, Badreddine Ahtchi, Austin Lewis, and Douglas Shontz, *Issues with Access to Acquisition Data and Information in the Department of Defense: Streamlining and Improving the Defense Acquisition Executive Summary (DAES) Process and Data*, Santa Monica, Calif.: RAND Corporation, 2018, Not available to the general public; Jeffrey A. Drezner, Megan McKernan, Austin Lewis, Ken Munson, Devon Hill, Jaime Hastings, Geoffrey McGovern, Marek Posard, and Jerry M. Sollinger, *Issues with Access to Acquisition Data and Information in the Department of Defense: Identification and Characterization of Data for Acquisition Category (ACAT) II–IV, Pre-MDAPs, and Defense Business Systems*, Santa Monica, Calif.: RAND Corporation, 2019, Not available to the general public.

For more information on the RAND Acquisition and Technology Policy Center, see www.rand.org/nsrd/ndri/centers/atp or contact the director (contact information is provided on the webpage).

Contents

Summary

Acquisition program data help drive effective and efficient policy formulation, decisionmaking, and program execution across the U.S. Department of Defense (DoD). Despite recent statutory changes to organizational structures, as well as to roles, responsibilities, and authorities of the Office of the Secretary of Defense (OSD) and military departments, OSD still needs key program data to inform policymaking and enable analysis that is critical for understanding acquisition processes and performance. For example, program data are still needed to conduct portfolio analyses and understand the performance of individual acquisition pathways and the overall acquisition system in order to improve acquisition policy design and outcomes.

This report outlines issues and opportunities in data requirements, governance, and management to strive for more efficient, effective, and informed acquisition while reducing burden and ad hoc data requests. We address general data governance and management challenges, as well as specific challenges associated with the Middle Tier of Acquisition for rapid prototyping and rapid fielding, the Selected Acquisition Report (SAR), and the Defense Acquisition Executive Summary (DAES) process and data. The acquisition community has a rich set of information at its disposal. Although there is no agreement on all data needs and definitions, the underlying data used for program management, oversight and insight, decisionmaking, and analysis are similar across DoD. Drawing on prior and ongoing research, we make several recommendations for using data to improve acquisition system performance:

- **Let decisionmaking drive data requirements**. Data must not be generated for their own sake but must support important decisionmaking about policy, process, programs, and integrated capability outcomes. As a starting point, we recommend that the Under Secretary of Defense for Acquisition and Sustainment (USD[A&S]) identify data requirements by specifying important acquisition use cases that must be supported across the department. Those use cases could be formalized along with data governance roles in an enterprise-wide acquisition data strategy.
- **Minimize reporting requirements and costs more generally**. For a given use case, we recommend that information and documentation requirements should be austere, with minimal data reporting. Historically, successful rapid prototyping and fielding activities have had austere information requirements. Guidance appears to recognize this by emphasizing tailoring.
- **Align and standardize where possible.** We recommend that emerging strategy, policy, and process for acquisition be supported by a common acquisition data framework. Ultimately, the data should align with the goals of the use cases that determine data requirements, promulgated in the enterprise-wide acquisition data strategy discussed above.

USD(A&S) should leverage the existing acquisition visibility data framework that reflects the legacy SAR and DAES data operations, provides a strong foundation from which to evolve, and has enabled rapid implementation of the Middle Tier acquisition pathway. We also recommend that DoD work with Congress to withdraw termination of the SAR until an appropriate substitute is developed.

- **Capitalize on existing structures**. One way to minimize costs and burdens (including ad hoc data calls) is by leveraging existing data frameworks, information systems, and organizations to the maximum extent practical, especially when such data are shared automatically between the Office of the USD(A&S) and component systems.

Consistent with OUSD(A&S) being an organization "with data-driven analysis linked to National Defense Strategy objectives,"[2] we recommend continued USD(A&S) attention to these and other acquisition data–related issues. Data enable improved policy formulation and decisionmaking across DoD, ultimately accelerating delivery of operational capabilities to the warfighter.

---

[2] Ellen Lord, "Department of Defense Press Briefing on DoD Acquisition Reforms and Major Programs," Office of the Under Secretary of Defense (Acquisition and Sustainment), May 10, 2019.

Acknowledgments

We would like to thank Mark Krzysko, Principal Deputy Director, Enterprise Information, within the Office of the USD(A&S) for thought-provoking discussions on data governance, management, and access in DoD throughout the study. We also thank Mark Hogenmiller, Laura Cooper, and Rhonda Edwards, who provided additional information in support of this research.

We are grateful to the formal peer reviewers of this document, Philip S. Antón, Irv Blickstein, and Obaid Younossi, who helped improve each short paper and the overall effort through their comments and suggestions. We also thank Maria Falvo for her assistance during this effort.

Finally, we would like to thank the director of the RAND Corporation's Acquisition and Technology Policy Center, Joel Predd, for his insightful comments on this research.

Abbreviations

ACAT	Acquisition Category
A&S	Acquisition and Sustainment
CMO	Chief Management Officer
DAE	Defense Acquisition Executive
DAES	Defense Acquisition Executive Summary
DAMIR	Defense Acquisition Management Information Retrieval
DASD	Deputy Assistant Secretary of Defense
DAVE	Defense Acquisition Visibility Environment
DCMA	Defense Contract Management Agency
DEPSECDEF	Deputy Secretary of Defense
DoD	U.S. Department of Defense
DoDD	Department of Defense Directive
FY	fiscal year
IT	information technology
JCIDS	Joint Capabilities Integration Development System
MDA	milestone decision authority
MDAP	major defense acquisition program
NDAA	National Defense Authorization Act
O&S	Operating and Support
OSD	Office of the Secretary of Defense
OUSD(A&S)	Office of the Under Secretary of Defense for Acquisition and Sustainment
OUSD(AT&L)	Office of the Under Secretary of Defense for Acquisition, Technology, and Logistics
PM	program manager
R&E	Research and Engineering

RRAs	roles, responsibilities, and authorities
SAE	Service Acquisition Executive
SAR	Selected Acquisition Report
SECDEF	Secretary of Defense
USD(A&S)	Under Secretary of Defense for Acquisition and Sustainment
USD(R&E)	Under Secretary of Defense for Research and Engineering

CHAPTER ONE

Introduction

Acquisition data lay the foundation for decisionmaking, program management, insight, and oversight of the U.S. Department of Defense's (DoD's) acquisition system. Recent statutory changes to organizational structures, as well as roles, responsibilities, and authorities (RRAs) have introduced new challenges and opportunities that will significantly affect the collection, storage, and use of acquisition information.

To help DoD and, particularly, the newly created Under Secretary of Defense for Acquisition and Sustainment (USD[A&S]) consider the implications of these changes and respond accordingly, we conducted research to identify and describe some of the issues and challenges associated with managing acquisition program information in the emerging acquisition environment. The short papers generated during this study, presented in this report as separate chapters, help inform some of the detailed decisions DoD must make as it implements the recent statutory changes.

The fiscal year (FY) 2016 and FY 2017 National Defense Authorization Acts (NDAAs) included significant changes in the organizational structure and RRAs of service and the Office of the Secretary of Defense (OSD) entities managing and overseeing acquisition programs. The Service Acquisition Executive (SAE) is now the default milestone decision authority (MDA) for new major defense acquisition programs, and the Service Chiefs have a reinforced role in contextual decisions of requirements, budgets, delivery requirements, and cost. Beginning in February 2018, the Office of the Under Secretary of Defense for Acquisition, Technology, and Logistics (OUSD[AT&L]) was dissolved, and two new under secretariats were created: one for Acquisition and Sustainment (A&S) and one for Research and Engineering (R&E). In addition, a Chief Management Officer (CMO) position was established in OSD and given responsibility for

> establishing policies on, and supervising, all business operations of the Department, including business transformation, business planning and processes, performance management, and business information technology (IT) management and improvement activities and programs, including the allocation of resources for business operations, and unifying business management efforts across the Department.[1]

There is potentially some overlap in acquisition program information management within DoD among these three positions (SAE, Service Chiefs, and CMO) and their accompanying organizations.

[1] Public Law 114-328, National Defense Authorization Act for Fiscal Year 2017, Section 901, December 23, 2016.

At the time of this writing (February 2019), the Services and OSD have implemented these changes but continue to work through some implementation details.[2] One area that will be significantly affected by these implementation decisions is acquisition program information. Program information is currently generated, collected, stored, accessed, and used by many organizations in the Services and OSD. Implementing these changes in organizational RRAs and structures will necessarily affect the generation, collection, storage, and use of acquisition data. In particular, it may hinder identifying the authoritative source of specific data, disrupt collection, and limit access and use. As responsibilities move to the Services, Service staff may need to develop new or expanded capabilities, particularly in terms of oversight function and capacity for major programs.

While these changes—the emerging acquisition environment—will have a significant impact on acquisition program data governance and management, acquisition program information will still be needed to support program management, analysis, and oversight.

Key Scoping Assumptions

Recent changes in acquisition RRAs prompted a debate about what acquisition data are required for the USD(A&S) to execute evolving acquisition responsibilities. Ultimately, acquisition program data requirements are the purview of USD(A&S) and depend on how USD(A&S) intends to use the data—i.e., the use cases—and on what financial costs and potential managerial and administrative burdens the department is willing to accept to collect, manage, store, share, and govern acquisition program data and information. This research did not address this basic question, which bears on broader questions of acquisition policy. Instead, we assume that USD(A&S) will continue to need acquisition program data to support a broad set of use cases. These use cases include the following:

- statutory and regulatory reporting
- tracking program cost, schedule, and performance outcomes against an established baseline
- providing program insight and oversight to anticipate, understand, and mitigate the factors affecting adverse cost, schedule, and performance outcomes
- conducting portfolio analyses, including both traditional (i.e., by Service or weapon system type) and new (i.e., mission-focused kill chains)
- understanding the performance of the overall acquisition system or any specific pathway within that system (e.g., traditional, tailorable Department of Defense Instruction 5000.02; Middle Tier) to improve policy design and implementation.

This assumption scopes our analysis, because ultimately, USD(A&S) can decide that some of these use cases (or their specific instantiations) are no longer needed in the new environment, or that the costs and potential burdens associated with collecting, managing, storing,

---

[2] DoD, *Report to Congress Restructuring the Department of Defense Acquisition, Technology and Logistics Organization and Chief Management Officer Organization, in Response to Section 901 of the National Defense Authorization Act for Fiscal Year 2017 (Public Law 114-328)*, Washington, D.C., August 2017. Organization charts issued since publication of this DoD response to Congress show some variation from the original proposed structure.

sharing, and governing acquisition program data cannot be justified. Analyses of such trade-offs are left for future work.

The topics listed above address only a few of the challenges associated with acquisition program data governance and management due to the Acquisition, Technology, and Logistics (AT&L) reorganization, MDA changes for major acquisition programs, and other changes in RRAs. Although not exhaustive, they provide a sample of challenges the department will need to confront in the emerging acquisition environment.

Research Objective and Approach

The objective of this research[3] was to identify and concisely describe some of the issues and challenges associated with managing acquisition program information in the emerging acquisition environment. The intent was to inform, in a timely manner, some of the policy design and implementation decisions DoD must make in response to the recent statutory changes to authorities, responsibilities, and organizational structure.

Our approach consisted of three main steps. First, the study team identified and described recent changes to DoD acquisition RRAs. This step was fundamentally descriptive in nature and was accomplished by reviewing relevant legislation and acquisition policy changes, and by interviewing DoD leadership in charge of developing policy to guide or implement the changes.

Second, the study team identified a set of specific challenges for acquisition data that may arise from the changes in RRAs. The topics were chosen with approval of the sponsor but were informed by six earlier studies on *Issues with Access to Acquisition Data and Information in the Department of Defense*.[4] Several topics were ultimately selected:

---

[3] This study was commissioned by Mark Krzysko, director, Acquisition Data, within the Office of the Under Secretary of Defense for Acquisition and Sustainment (OUSD[A&S]). This report draws on that larger study, as well as on prior research on acquisition program data and the authors' experience as acquisition researchers.

[4] Jessie Riposo, Megan McKernan, Jeffrey A. Drezner, Geoffrey McGovern, Daniel Tremblay, Jason Kumar, and Jerry M. Sollinger, *Issues with Access to Acquisition Data and Information in the Department of Defense: Policy and Practice*, Santa Monica, Calif.: RAND Corporation, RR-880-OSD, 2015; Megan McKernan, Jessie Riposo, Jeffrey A. Drezner, Geoffrey McGovern, Douglas Shontz, and Clifford A. Grammich, *Issues with Access to Acquisition Data and Information in the Department of Defense: A Closer Look at the Origins and Implementation of Controlled Unclassified Information Labels and Security Policy*, Santa Monica, Calif.: RAND Corporation, RR-1476-OSD, 2016; Megan McKernan, Nancy Young Moore, Kathryn Connor, Mary E. Chenoweth, Jeffrey A. Drezner, James Dryden, Clifford A. Grammich, Judith D. Mele, Walter T. Nelson, Rebeca Orrie, Douglas Shontz, and Anita Szafran, *Issues with Access to Acquisition Data and Information in the Department of Defense: Doing Data Right in Weapon System Acquisition*, Santa Monica, Calif.: RAND Corporation, RR-1534-OSD, 2017; Megan McKernan, Jessie Riposo, Geoffrey McGovern, Douglas Shontz, and Badreddine Ahtchi, *Issues with Access to Acquisition Data and Information in the Department of Defense: Considerations for Implementing the Controlled Unclassified Information Reform Program*, Santa Monica, Calif.: RAND Corporation, RR-2221-OSD, 2018; Jeffrey A. Drezner, Megan McKernan, Badreddine Ahtchi, Austin Lewis, and Douglas Shontz, *Issues with Access to Acquisition Data and Information in the Department of Defense: Streamlining and Improving the Defense Acquisition Executive Summary (DAES) Process and Data*, Santa Monica, Calif.: RAND Corporation, 2018, Not available to the general public; Drezner, Jeffrey A., Megan McKernan, Austin Lewis, Ken Munson, Devon Hill, Jaime Hastings, Geoffrey McGovern, Marek Posard, and Jerry M. Sollinger, *Issues with Access to Acquisition Data and Information in the Department of Defense: Identification and Characterization of Data for Acquisition Category (ACAT) II–IV, Pre-MDAPs, and Defense Business Systems*, Santa Monica, Calif.: RAND Corporation, 2019, Not available to the general public.

- general data governance and management issues associated with the emerging acquisition environment
- specific data challenges associated with the implementation of the Middle Tier acquisition pathway
- implications of termination of the Selected Acquisition Report (SAR)
- future of the Defense Acquisition Executive Summary (DAES) process and associated data.

Third, the study team identified implications for acquisition data for each of the selected topics. These implications were developed on the basis of published best practices for data management and an understanding of how those practices are currently implemented in the DoD acquisition system. Where possible, the study team also identified how current DoD policies and practices may need to change to be consistent with the emerging and future acquisition environment (in terms of roles, responsibilities, and structure) and identified options for mitigating the challenges.

Organization of This Report

The research generated four short papers, which are presented here as chapters. Chapter Two focuses on general data governance and management issues associated with the emerging acquisition environment. Chapter Three provides data concerns associated with the implementation of the Middle Tier acquisition pathway. Chapter Four discusses the termination of the SAR. Chapter Five examines the future of the DAES process and associated data. Finally, Chapter Six presents a set of key observations and potential options for DoD to consider going forward for each of these topic areas.

CHAPTER TWO

Opportunities for Improved Acquisition Information Management in the Emerging Acquisition Environment

Acquisition information is critical to the efficient and effective operation of the defense acquisition system. This information is fundamental to those in decisionmaking, management, execution, insight, and oversight roles. We identified challenges to the collection, storage, and use of acquisition information, as well as options for improving access to and management of these data.[1]

As with any large, complex organization, DoD faces challenges related to data access and management. Recent statutory changes to organizational structures and RRAs will exacerbate these challenges and have introduced new challenges and opportunities that will significantly affect the collection, storage, and use of acquisition information. The USD(A&S) can play a critical role in both tackling challenges and leveraging opportunities through policy and guidance on implementing the statutory changes.

Before the current reorganization and statutory changes, challenges affecting acquisition information included complex security policies regulating information systems; cultural and technical barriers to accessing and sharing information; and lack of awareness of the breadth and depth of information available to DoD leaders and staff. A rich set of information is available to support acquisition insight and decisionmaking across myriad central information systems, but the full extent to which this information is used remains unknown. In addition, no common data environment exists for all acquisition information, there is no agreement on all data needs and definitions across DoD, and DoD leadership has not explored or supported this concept over time. These issues result from decentralized governance and management. While most information used for program management and oversight and insight is similar across OSD and the Services (at least for similarly sized programs), specific performance metrics and uses differ. Finally, introducing changes to rules regarding controlled unclassified information will further complicate management, sharing, and use of acquisition information in the near term.[2]

---

[1] The deputy director, Acquisition Resources and Analysis, Enterprise Information within the OUSD(A&S) sponsored this research.

[2] Riposo et al., 2015; McKernan et al., 2016; McKernan et al., 2017; Drezner et al., 2018; Drezner et al., 2019; McKernan et al., 2018.

Congressional Changes Have Resulted in an Uncertain Environment for Acquisition Information Management and Use

Recent NDAAs have significantly changed defense acquisition roles, responsibilities, and organizational structure within DoD. Section 825 of the FY 2016 NDAA delegated decisionmaking to the SAEs for new major defense acquisition programs (MDAPs),[3] while Section 901 of the FY 2017 NDAA eliminated the Under Secretary of Defense for Acquisition, Technology and Logistics position and created an Under Secretary of Defense for Research and Engineering (USD[R&E]) and the new USD(A&S), along with a CMO.[4] In addition to these major leadership changes with associated supporting organizational changes, some legislation directly affects data management. For instance, section 913 of the FY 2018 NDAA requires the Secretary of Defense (SECDEF) to establish "activities that use data analysis, measurement, and other evaluation-related methods to improve . . . acquisition program outcomes."[5] Also, section 1654 of the FY 2018 NDAA requires status reporting of Nuclear Command, Control, and Communications (NC3) acquisition programs,[6] as well as requiring that the Chief Information Officer create a database relating to the execution of all NC3 acquisition programs with an approved Materiel Development Decision. The current absence of data governance over such information creates uncertainties regarding the authoritative sources, definitions, and standards of specific data, impeding collection and limiting access and use. As decision and oversight responsibilities move to the military departments, their staffs may need to develop new or expanded capabilities.

Statutory Changes Significantly Affect Information Flows in the Department of Defense

Implementing these changes in organizational roles, responsibilities, and structures will affect the generation, collection, storage, and use of acquisition data. The USD(A&S) plays a critical role in determining what acquisition information is needed across the DoD enterprise (in particular, outside the military departments). These organizational changes present an opportunity to redefine what acquisition information different organizations need and how that information is collected, stored, protected, accessed, and shared.

Key questions senior acquisition leaders need to consider include the following:

- What information does OSD and the Fourth Estate need and why? In particular, what does USD(A&S) need (e.g., to execute her statutory responsibility to advise the SAEs on acquisition decisions, to inform policymaking, to inform the SECDEF and Deputy Secretary of Defense [DEPSECDEF], for mission and kill-chain portfolio views, and for congressional reporting, cognizance, and oversight)?
- Is it possible to have decentralized program execution and oversight while maintaining OSD insight on policy effects, institutional performance, and key program status and

[3] See Public Law 114-92, National Defense Authorization Act for Fiscal Year 2016, November 25, 2015.

[4] Pub. L. 114-328, 2016.

[5] Public Law 115-91, National Defense Authorization Act for Fiscal Year 2018, December 12, 2017.

[6] See Pub. L. 115-91, 2017.

outcomes? What are the resulting OSD requirements for critical, high-value data and information?
- How will mission and kill-chain portfolio performance and program interdependencies be monitored and improved in this decentralized structure?
- How can data and enabled analysis improve the execution of programs?
- What data and analytic capabilities and insights will be lost if some information flows stop? What net savings or costs would ensue?
- What information is no longer needed or of low value?
- What new high-value information is needed?
- Which acquisition program data should be standardized across the DoD enterprise and across different Services and types of programs?
- What are the military departments doing with their information flows as their organizations change?

Options to Consider

To mitigate any unintended consequences from these recent statutory changes, USD(A&S) could address these challenges through policy and guidance, as well as organizational restructuring. For example,

- USD(A&S) could create a strategic management plan for acquisition information that determines what information is needed by whom to accomplish enterprise-wide objectives without overburdening the military departments. This near-term guidance could help keep the required data flows from atrophying.
- USD(A&S) and the military department leadership could work together to standardize a core set of data definitions, authoritative sources, and management approaches. This effort would facilitate information sharing and understanding and would be an important first step toward a common acquisition data framework.

CHAPTER THREE

Implications of the Middle Tier Acquisition Pathway for the Management, Sharing, and Governance of Acquisition Data

Creating a Middle Tier acquisition pathway for programs that emphasize accelerated capability development and deployment through rapid prototyping and rapid fielding both poses challenges and offers DoD opportunities. Such a pathway will require some acquisition information to manage, execute, and oversee the process and its content. Those managing acquisitions along this pathway also will need information (and processes) to support decisions on initiation, budgeting, requirements, transition, production, and fielding and deployment. This chapter describes several implications of the Middle Tier pathway for the governance and management of those data.

The Middle Tier Acquisition Pathway

Congress directed DoD to reemphasize the use of prototyping by establishing new authorities.[1] Section 804 of the FY 2016 NDAA (as amended) directs the creation of a Middle Tier of Acquisition for Rapid Prototyping and Rapid Fielding.[2] For rapid prototyping, the objective "shall be to field a prototype that can be demonstrated in an operational environment and provide for a residual operational capability within five years of the development of an approved requirement." DoD is also directed to develop "a process for transitioning successful prototypes to new or existing acquisition programs for production and fielding under the rapid fielding pathway or the traditional acquisition system." The Middle Tier acquisition pathway is ". . . distinct from the traditional acquisition system. Under the Middle Tier of Acquisition, programs subject to the guidance shall not be subject to the Joint Capabilities Integration Development System (JCIDS) manual and DoD Directive 5000.01, 'The Defense Acquisition System,' except to the extent specifically provided in the implementing guidance."[3]

Section 806 of the FY 2017 NDAA establishes additional processes and reporting on prototyping projects. It grants each SAE authority to select prototype projects that will be

[1] The brief highlights in this section are current as of U.S. Code, Title 10, Armed Forces, release point 115-137, March 16, 2018.

[2] See "Middle Tier of Acquisition for Rapid Prototyping and Rapid Fielding" in the Historical and Revision Notes after Section 2302 (Definitions) in the 115-137 release point of 10 U.S.C.

[3] USD(A&S), "Middle Tier of Acquisition (Rapid Prototyping/Rapid Fielding) Interim Authority and Guidance," memorandum, Washington, D.C.: U.S. Department of Defense, April 16, 2018a, p. 1; USD(A&S), "Middle Tier of Acquisition (Rapid Prototyping/Rapid Fielding) Interim Guidance," memorandum, Washington, D.C.: U.S. Department of Defense, October 9, 2018b; USD(A&S), "Middle Tier of Acquisition (Rapid Prototyping/Rapid Fielding) Interim Guidance 2," memorandum, Washington, D.C.: U.S. Department of Defense, March 20, 2019.

completed within two years, cost up to $10 million (or up to $50 million with approval by the Secretary of the military department or the Secretary's designee) in FY 2017 dollars, and be subject to certain reporting and oversight requirements.[4] Regarding data, U.S.C. Section 2447a levies new budgetary reporting requirements for advanced component development and prototyping activities.[5]

Interim guidance from the USD(A&S) provided parameters regarding information requirements for the Middle Tier acquisition pathway.[6] It also identified an initial set of core information that should be collected regarding these efforts (at a minimum) and discussed a data-driven collaborative policymaking process drawing on lessons learned from the initial implementation. The Navy and the Air Force released guidance in April 2018, and the Air Force followed up with additional detailed guidance in June.[7] One area of similarity between the Navy and Air Force guidance is the emphasis on tailoring current statutory and regulatory information requirements and seeking waivers as needed to minimize information requirements and help maintain schedule, making tailoring a key tool that program managers (PMs) will need to use. The tailoring will be driven by the unique characteristics of the efforts and to the decisions being made by the MDA.

Challenges for Information Management

A preliminary analysis of recent Middle Tier acquisition pathway guidance suggests it will face data-related challenges. We identify four of those challenges below.

---

[4] U.S. Code, Title 10, Section 2447a–c, Weapon System Component or Technology.

[5] 10 U.S.C. 2447a, Weapon system component or technology prototype projects: display of budget information:

> (a) Requirements for Budget Display. In the defense budget materials for any fiscal year after fiscal year 2017, the Secretary of Defense shall, with respect to advanced component development and prototype activities (within the research, development, test, and evaluation budget), set forth the amounts requested for each of the following:
>
> (1) Acquisition programs of record.
> (2) Development, prototyping, and experimentation of weapon system components or other technologies, including those based on commercial items and technologies, separate from acquisition programs of record.
> (3) Other budget line items as determined by the Secretary of Defense.
>
> (b) Additional Requirements. For purposes of subsection (a)(2), the amounts requested for development, prototyping, and experimentation of weapon system components or other technologies shall be—
>
> (1) structured into either capability, weapon system component, or technology portfolios that reflect the priority areas for prototype projects; and
> (2) justified with general descriptions of the types of capability areas and technologies being funded or expected to be funded during the fiscal year concerned.

[6] USD(A&S), 2018a, p. 3.

[7] Assistant Secretary of the Navy for Research, Development, and Acquisition, "Middle Tier Acquisition and Acquisition Agility Guidance," memorandum, Washington, D.C.: Department of the Navy, April 24, 2018, pp. 1–3; Assistant Secretary of the Air Force for Acquisition, Technology and Logistics, "Seven Steps for Incorporating Rapid Prototyping into Acquisition," memorandum, Washington, D.C.: Department of the Air Force, April 10, 2018a; Assistant Secretary of the Air Force for Acquisition, Technology, and Logistics, "Air Force Guidance Memorandum for Rapid Acquisition Activities," memorandum, Washington, D.C.: Department of the Air Force, June 13, 2018b, pp. 7–8.

Note: When the original short paper was released (June 2018), the Assistant Secretary of the Army for Acquisition, Logistics and Technology had not yet released guidance related to the Middle Tier acquisition pathway or rapid prototyping generally; that guidance was formally released in September 2018.

Challenge 1: Volume of Information

Recently released implementation guidance from the Air Force and the Navy describes a significant amount of information that will be generated and collected.

Some examples from the Air Force guidance include the following:

- It is the responsibility of the initiating or assigned PM to propose required program documentation, decision points, metrics, guardrails, as well as timing and scope of decision reviews, and to establish cost, schedule, risk, and performance objectives. This strategy should be determined at program initiation, approved by the MDA, and documented in the [Acquisition Strategy Document] . . .
- . . . The PM should provide adequate information to support Air Force evaluation of cost, schedule, and performance and to support MDA, Office of the Secretary of Defense, and Congressional reporting where required . . .
- . . . The PM shall provide a brief (e.g., 1- to 2- page) tri-yearly report summarizing status of progress towards cost, schedule, and performance objectives; progress toward design goals; and likelihood of crossing guardrail thresholds in future.[8]

The Navy has also provided some specific guidance on information requirements. Examples include the following:

- The PM should review the documentation in Table 2-1 and make recommendations to the Approval Authority to identify which documents best support project execution when utilizing the Rapid Prototyping pathway . . .
- The PM should review the documentation in Table 2-2 and 2-3 and make recommendations to the Approval Authority to identify which documents will best support program execution when utilizing the Rapid Fielding pathway. In order to optimize production and fielding timelines, the PM may tailor timelines for developing program documents that are not required for program initialization. The documentation tailoring plan shall be included in [the] Rapid Fielding plan.[9]

The tables referenced in the Navy guidance list documentation required by statute or regulation under the traditional acquisition process; therefore, it should be appropriate to use tailoring. In both cases, the data must be approved by the MDA (now the SAE). This guidance appears to be in the spirit of the legislation, in that data are not a one-size-fits-all commodity, and both Services are attempting to avoid collecting data only for that purpose; however, this guidance may naturally lead to the end of any continuity or regularity of data collection on Middle Tier programs.

Challenge 2: Uncoordinated Reporting Requirements

At the PM and MDA levels, the guidance on what data need to be collected and used for decisionmaking and execution appears to be extensive; the possibility of significant information management challenges is likely without a more focused DoD-wide plan. The guidance released by OSD, Air Force, and Navy does not coordinate information requirements, the vast

---

[8] Assistant Secretary of the Air Force for Acquisition, Technology, and Logistics, 2018b, pp. 1, 7, 9.

[9] Assistant Secretary of the Navy for Research, Development, and Acquisition, 2018, p. 2.

amount of information that the Middle Tier acquisition pathway will likely generate, or the terminology being used. In particular, this new pathway for acquisition has significant congressional interest that requires a DoD-wide answer for how the organization is collectively implementing this new pathway and what benefits or challenges have been experienced during implementation. Such an effort requires common terminology to be able to compare across the organization.

In addition, the Air Force appears to deviate from common acquisition terminology. Air Force guidance identified "phases" for rapid fielding, a term that is not typically used in acquisition. The Air Force is permitted to create policy that fits its organization; however, any differences in the terminology used for the Middle Tier acquisition pathway will likely result in challenges that will need to be thought through by those managing the output. For example:

> Section 804 should be used for all future Air Force acquisitions to the maximum extent practicable when deemed suitable by the MDA. This acquisition approach has up to four tailorable phases depending on the pathway chosen:
>
> - 3.4.1. Alpha: Prototyping,
> - 3.4.2. Beta: Fielding and Initial Production,
> - 3.4.3. Gamma: Modernization and Follow-on Production,
> - 3.4.4. Delta: Operations and Sustainment.[10]

The label for each of these phases may cause confusion; Alpha, Beta, Gamma, and Delta are not labels normally associated with any acquisition pathway. These labels, as well as their associated descriptions, do not clearly identify the relationship among them or their relationship to the traditional acquisition phases.

Challenge 3: Lack of Standardization

The USD(A&S), the USD(R&E), and the military departments must also consider who will be responsible for governing and managing the information to avert some additional challenges already visible in DoD-wide acquisition program management. For instance, current policy has disparate guidance for identifying when efforts start, when they become an *acquisition program*, and how an *acquisition program* is defined. In other words, current guidance does not establish a common way to assign a "birth certificate" to acquisition programs across DoD. This dynamic will also affect the Middle Tier acquisition pathway due to the following guidance on *initiation* that differs among Services:

Navy:

> - Initiation. The Middle Tier Acquisition pathways require rapid prototyping projects and rapid fielding programs to be completed within five (5) years of initiation. For rapid fielding, production must be scheduled to begin within six (6) months of initiation. Each pathway requires a distinct merit-based assessment to ensure the transparency, accountability and alignment within the Navy and Marine Corps acquisition, requirements, and resourcing communities. [Assistant Secretary of the Navy for Research, Development and Acquisition] will initiate all Middle Tier Acquisition programs by an Acquisition Decision Memorandum and may designate Acquisition

[10] Assistant Secretary of the Air Force for Acquisition, Technology, and Logistics, 2018b, p. 6.

Decision Authority at that time. A Program Manager will be designated and report directly to Acquisition Decision Authority.[11]

Air Force:

- The Beta Decision is of paramount importance. The MDA and PM will ensure prototyping efforts have aggressively tackled all cost-, schedule-, and performance-driving risks and conducted an operationally-relevant demonstration prior to initiation of a rapid fielding effort. This decision must terminate technically-flawed, prohibitively-costly, or operationally-unwieldy concepts with extreme prejudice. . . . Unlike traditional acquisition, rapid acquisition activities will not be considered Programs of Record until entering the Beta Phase.[12]

Although defining programs separately within each service may be satisfactory, these differences cause many issues for the industrial sector that works across all of the Services and OSD generally, including the Fourth Estate. Defining terms like *program* differently across DoD leads to inefficiency in trying to translate and piece together disparate definitions when the need arises to compare across the department.

Challenge 4: Overburdening the Unburdened Path

The intent of the Middle Tier acquisition pathway is to promote another means for DoD to increase the speed of acquisition. Congress eliminated the need to use JCIDS and Department of Defense Directive (DoDD) 5000.01, but a level of statutory and regulatory requirements remains that will need to be considered and waived if unnecessary along with Service-level oversight needs. For instance, the Navy provides the following guidance:

- Program Managers need to review the following documentation and make recommendations to the Approval Authority to identify which documents best support project execution when utilizing the Rapid Prototyping pathway:
  - Acquisition Decision Memorandum
  - Prototyping Plan: Prototype Acquisition Strategy, Performance Goals, Knowledge Points, System Experimentation and Assessment Plan, Prototype Deployment Strategy
  - Acquisition Plan (when required by the Federal Acquisition Regulation [FAR])
  - Capability Documentation (within 6 months of initiation): Top Level Requirements Document -or- [Joint Emergent Operational Needs], [Joint Urgent Operational Needs], Service [Operational Needs Statement] -or- [Capability Development Document] -or- [Capability Production Document]
- PMs should also review the documentation below and make recommendations to the Approval Authority to identify which documents will best support program execution when utilizing the Rapid Fielding pathway
  - Rapid Fielding Documentation – Statutory:
    - Acquisition Strategy: Acquisition Approach, Benefit Analysis and Determination, Business Strategy, Contracting Strategy (including Contract Type), Cooperative Opportunities, General Equipment Valuation, Industrial Base Capability Considerations (not to exceed 20 pages in length)

[11] Assistant Secretary of the Navy for Research, Development, and Acquisition, 2018, p. 4.

[12] Assistant Secretary of the Air Force for Acquisition, Technology and Logistics, 2018b, p. 6.

  - Acquisition Plan: When required by the Federal Acquisition Regulation (FAR). May be combined with Acquisition Strategy
  - Core Log Determination / Workload Estimate: May be summarized in acquisition strategy
  - Frequency Allocation: For all systems/equipment that use the electromagnetic spectrum while operating in the United States and its possessions
  - Low-Rate Initial Production Quantity: Production Quantities will be addressed in Acquisition Decision Memorandum
  - Post Implementation Review: Fulfilled by disposition decision to sustain Rapid Fielding program or transition to Program of Record
  - Cyber Security Strategy: Statutory for only mission critical or mission essential [information technology (IT)] systems
  - Clinger Cohen Act Compliance: Statutory for all programs that acquire IT systems
  - [Programmatic Environment, Safety and Occupational Health Evaluation] and [National Environmental Policy Act/Executive Order] 12114 Compliance Schedule: Not required for software programs with no hardware components
- Rapid Fielding Documentation – Regulatory:
  - Acquisition Decision Memorandum: Acquisition Decision, Program Cost Estimate, Rapid Fielding Quantities (fulfills Low-Rate Initial Production Quantity statute), Schedule
  - IT Deployment Strategy: Address applicable elements: IT & [National Security System] Interoperability Cert, Spectrum Supportability Risk Assessment, Bandwidth Requirement Review, Cyber Security Strategy (non–mission critical or mission essential IT systems), Program Protection Plan, Waveform Assessment Application
  - Capability Documentation (within 6 months of initiation): Top Level Requirements Document -or- [Joint Emergent Operational Needs], [Joint Urgent Operational Needs], Service [Operational Needs Statement] -or- [Capability Development Document] -or- [Capability Production Document]
  - Concept of Operations (CONOPS)
  - Course of Action Analysis: Replaces and serves as [Analysis of Alternatives]
  - Defense Intel Threat Library
  - Operational Test Plan
  - Operational Test Report
  - Systems Engineering Strategy
  - Sustainment Strategy
  - Training Systems Strategy[13]

The abundance of statutory, regulatory, and Service-level oversight needs may become one of several key challenges for those trying to meet the five-year goal for these efforts. Although the solution of "tailoring" is provided in the Air Force and Navy guidance, in practice, tailoring is challenging:

---

[13] This list is derived from information in table format within Assistant Secretary of the Navy for Research, Development, and Acquisition, 2018, p. 13.

> Tailoring recognizes that acquisition programs are not all the same, and policy permits and encourages program managers (PMs) to customize regulatory-based reviews, processes, and information requirements to accommodate the unique characteristics of a program while still meeting the regulations' intent for appropriate decision criteria and oversight processes. The extent to which programs take advantage of opportunities to tailor processes and documentation is not clear, but anecdotal evidence suggests that tailoring is more difficult in practice than guidance suggests.[14]

Likewise, additional research suggests that austere (i.e., minimal) documentation requirements are beneficial in prototyping efforts:

> . . . [Research] suggest[s] some of the conditions under which prototyping strategies are most likely to yield benefits in a development program. These conditions include ensuring that prototyping strategies and documentation are austere, not committing to production during the prototyping phase, making few significant design changes when moving to the final configuration and maintaining strict funding limits.[15]

Guidelines Moving Forward

The examples above highlight some of the data-related challenges that may arise during implementation and should be addressed in Middle Tier policy and guidance as implementation proceeds. One opportunity to address these challenges will be the collaborative policymaking process directed in the USD(A&S) interim guidance memorandum. Using prior research, we offer four guidelines to ensure that requirements and processes associated with Middle Tier program data and other acquisition information are as efficient and effective as possible.

Guideline #1: Let Decisionmaking Drive Data Requirements

The specific decisions associated with the Middle Tier pathway should weigh heavily in determining data needs. Data and information must not be generated for their own sake but must support important decisionmaking about policy, process, programs, and integrated capability outcomes. These decisions include merit-based project selection, program initiation, determination and validation of requirements, execution status and oversight, and transitions (i.e., field residual capability, merge into existing program, initiate new program, or terminate). Although many data items could be addressed, the requirements should focus on data required to inform decisions. Analysis of what data are needed, why, and the associated costs needs to be conducted—and soon. However, it must be noted that decisions at the service level for MDAPs may be quite different across the Services and different from oversight by the USD(A&S). Therefore, the defense industrial base is going to have to attempt to have its internal systems aligned not only with different military Services but also, now, with different program management styles.

---

[14] Megan McKernan, Jeffrey A. Drezner, and Jerry M. Sollinger, *Tailoring the Acquisition Process in the U.S. Department of Defense*, Santa Monica, Calif.: RAND Corporation, RR-966-OSD, 2015, p. viii.

[15] Jeffrey A. Drezner and Meilinda Huang, *On Prototyping: Lessons from RAND Research*, Santa Monica, Calif.: RAND Corporation, OP-267-OSD, 2009, p. 1.

Guideline #2: Minimize Reporting Requirements

Information and documentation requirements should be austere, with minimal data reporting, and explicit, so that the acquisition community is aware of what leadership needs and why. Historically, successful rapid prototyping and fielding activities have had austere information requirements. Guidance appears to recognize this need by emphasizing tailoring through waivers; however, past research suggests that obtaining waivers can require significant effort.[16] It is important to remember that rapid prototyping activities are not necessarily programs, so imposing the full bureaucratic structure of program oversight and reporting can greatly slow activities, which is probably what motivated Congress' desire to bypass JCIDS and the traditional DoDD 5000.01 acquisition processes. Likewise, rapid fielding may be more like procurement than like full-blown acquisition programs. Costs (in bureaucratic terms, dollars, and in slowing the prototyping and fielding) must be considered in concert with the decision benefits. To trim reporting requirements and maintain some standards, the continuity and consistency does not have to cover all aspects of defense acquisition but should cover a core set that the acquisition community uses (e.g., cost, schedule, performance).

Guideline #3: Standardize Where Possible

A common framework should be developed for a core set of program data. This step would standardize data elements and definitions across DoD components while designating authoritative sources of the information; facilitating communication among OSD, the components, Congress, and external organizations (e.g., U.S. Government Accountability Office, industry); and enabling strategic-level analyses. This framework would provide answers to the following key questions regarding governance and management of the information that are missing from current guidance:

- What core information should be collected, standardized, and defined for a DoD-wide (i.e., an enterprise-wide) view of Middle Tier acquisition?
- What is the authoritative source for all facets of the information generated from the Middle Tier acquisition pathway?
- Where will the data be collected? Should a current information system be used, or should an additional information system be created?
- What is the plan for seamlessly sharing and promoting collaboration using this information so that DoD is inputting data once and sharing across many uses and organizations?
- Who will govern this information?
- Who is the data owner?
- Who is the information manager?
- Who owns the data definitions for the core information that is collected?
- Who can waive data reporting and under what circumstances?

In the following interim guidance, USD(A&S) has started to lay the groundwork for coordination, but more needs to be done:

> - Organizations utilizing this interim authority must also identify data that can be shared across the Department via an open and collaborative Department-managed

---

[16] McKernan, Drezner, and Sollinger, 2015.

tool under the final guidance. This data must include, but not be limited to: name of program, capability gap or problem, definitive source for the capability gap or problem, capability characteristic or solution, date funds approved for initiation, funding source, program result (transition or termination), date of transition or termination, reason for transition or termination, program budget, and vendor name(s) . . .

- . . . The [Component Acquisition Executive], for organizations utilizing this authority, is responsible for capturing and storing the above data. This data will be provided during the collaborative policy development effort . . . .[17]

One critical element for the new Middle Tier acquisition pathway is the ability to facilitate technology transition. Ideally, the information being collected for prototyping could be similar to acquisition program-of-record data, which will facilitate technology transition.

Guideline #4: Capitalize on Existing Structures

Existing data frameworks, information systems, and organizations should be used to the maximum extent practical. A centralized location for information storage might also be considered, using methods already in place to manage and share program information.

- The Air Force currently uses Project Management Resource Tools to input, store, and share its Acquisition Category (ACAT) I–III acquisition program information.
- The Navy uses the Research, Development, and Acquisition Information System to store its ACAT I–IV program information.
- The Army uses a combination of briefings and OSD-level information systems for its program information.
- USD(A&S) has four authoritative information systems to store ACAT I program information, ACAT I–approved acquisition documentation, and ACAT II authoritative program lists:
  - Defense Acquisition Management Information Retrieval (DAMIR)
  - Acquisition Information Repository
  - Defense Acquisition Visibility Environment (DAVE)
  - Earned Value Management Central Repository.

The DoD Comptroller is developing new systems for improved budgetary data collection and specificity. The Services and OSD have spent years developing solutions to be able to share this information electronically (e.g., using application programming interfaces), so this preexisting architecture should provide immediate benefits.

A strategic data management and governance plan for the Middle Tier pathway that follows these guidelines will facilitate successful implementation and enhance integration of the Middle Tier pathway with other acquisition paths throughout the DoD enterprise.

---

[17] USD(A&S), 2018a, p. 3.

CHAPTER FOUR

Implications of the Reform of the Selected Acquisition Report for the Governance, Management, and Sharing of Acquisition Data

A recent and somewhat obscure statutory change related to acquisition management and oversight requires senior DoD leadership attention: the forthcoming end of the requirement to submit SARs to Congress. The SAR has been a bedrock of transparency and data on the cost, schedule, and performance of MDAPs for oversight and analysis at the program, portfolio, and policy levels—both immediately and longitudinally. Such analyses have been critical to improving and informing weapon system acquisition strategies and policymaking in DoD and Congress for decades.[1]

The submission of a SAR for each major acquisition program to Congress was repealed by the FY 2018 NDAA effective December 31, 2021.[2] While this change is part of Congress' broader effort to ease DoD's reporting burden, RAND Corporation researcher discussions with professional staff members in the Senate and House Armed Services Committees indicate that Congress' intent was for DoD to review and propose a revised reporting structure that satisfies Congress' need for detailed, transparent performance information, but in a way that DoD finds more efficient and effective. The paper reproduced in this chapter was requested by our DoD sponsor to ensure that Congress' message and intent are not lost on DoD leadership and to begin outlining some key considerations for DoD's possible replacement or revision of the SAR. At the very least, this effort is particularly important because SARs have become critical informational elements for DoD's own insights into the effectiveness of policies, processes, management, and performance of the defense acquisition system.

Consequences of Terminating the Selected Acquisition Reports

The SAR has been used since 1968—predominantly to understand the cost, schedule, and performance of MDAPs.[3] SARs are important because, collectively, they provide a structured and relatively consistent mechanism for informing Congress about the performance of major investments; have been used for decades by DoD and its components for management and oversight; describe performance in a consistent way that is understood by all; and have been

[1] Joseph G. Bolten, Robert S. Leonard, Mark V. Arena, Obaid Younossi, and Jerry M. Sollinger, *Sources of Weapon System Cost Growth: Analysis of 35 Major Defense Acquisition Programs*, Santa Monica, Calif.: RAND Corporation, MG-670-AF, 2008.

[2] See Pub. L. 115-91, 2017, Section 1051(x)(4).

[3] Comptroller General of the United States, "Response to the Honorable F. Edward Herbert, Chairman, Committee on Armed Services, House of Representatives," B-163058, October 30, 1973, p. 2.

one of the only sources of longitudinal, standardized program information supporting program, portfolio, process, and policy analysis for the almost $100 billion[4] in annual investment in MDAPs.[5] The structure and data definitions for a SAR also constitute an important starting place for the development of common acquisition program data management across all program levels, program types, and components.

Elimination of this information source by Congress will, in turn, eliminate many of the benefits that have accrued from its use over time. Of particular concern is the potential loss of common data standards and definitions for measuring program performance. Without these common data standards and definitions (i.e., a common data framework), institutionalized over decades of SAR creation and submission, the military departments' performance measurements (i.e., definitional standards) may drift over time, leading to significant confusion and inefficiencies, a reduced ability to integrate business practices across DoD, and reduced transparency.

Moreover, the duties of the USD(A&S), as codified in Section 133b of U.S.C. Title 10, require the type of information collected and maintained in the SARs to inform policymaking and to elucidate portfolio issues. In particular, the USD(A&S)'s role of chief acquisition and sustainment officer—acting as the principal adviser to SECDEF on acquisition—involves establishing policies that enable an efficient and effective acquisition process. In principle, policymaking requires the kind of information contained in the SAR.

Opportunities to Improve the Selected Acquisition Report While Retaining Key Elements

Some data elements of the SAR are critical for information or analytic purposes, while others could be improved, streamlined, or eliminated. Below are some example opportunities, core elements, and hidden needs. These examples highlight known uses that further analysis could refine to improve the usefulness of SAR-like reporting to multiple user communities.

- **Streamlining and integration with other information sources.** DoD could review and integrate sources of similar information to reduce burden and increase efficiency rather than create pieces solely for the SAR. For example:
  - *Executive summary.* This is often the key authoritative source of program history, status, purpose, and plans. These summaries have been used frequently in analyses to understand overall program history from Milestone B through Milestone C. Potentially, this information could be sourced from or integrated with other similar sources (e.g., descriptions in the Comptroller's annual budget report on "Program Acquisition Cost by Weapon System" or the Director, Operational Test and Evaluation's annual reports) to reduce overall burden.

[4] The President's budget request for FY 2019 included $92.3 billion for MDAPs (see Office of the Under Secretary of Defense [Comptroller]/Chief Financial Officer, *Program Acquisition Cost by Weapon System, Fiscal Year 2019 Budget Request*, Washington, D.C.: U.S. Department of Defense, February 2018).

[5] Comptroller General of the United States, 1973, p. 2. See also OUSD(A&S), "Background of SAR," Washington, D.C.: U.S. Department of Defense, accessed via DAMIR, as of October 3, 2018.

- **Revision of certain elements.** Some elements are known to be problematic, particularly burdensome, or of little value, while others are valuable but require significant additional work to prepare. For example:
  - *Cost variance.* This section is known to have significant theoretical issues in how cost change types are allocated to statutory bins. However, some of this information has been useful for helping inform analysis of major trends in cost variance and root cause analyses. The textual descriptions can be somewhat informative, but the bins often overlap, making the binning subjective and, thus, much less informative. Here, DoD could develop and propose an alternative approach that is less confusing and more informative.
  - *Operating and support (O&S) cost.* While cost information is potentially valuable to those who are seeking to understand high-level O&S costs, it is difficult to understand how these costs, their uncertainties, and changes over time are due to external factors beyond the control of the acquisition system (e.g., whether the system is operated in an environment it was not designed for, whether operational tempo is increased). Possible improvements might include adding data on elements that drive sustainment costs (e.g., more consistently including reliability information and enriching the information on maintainability). Also, it may be useful to provide actual O&S costs if the program has operational units while still in production and reporting.
- **Key elements for retention (including some that seem obscure and burdensome).** Some elements are very valuable, but their value may not be apparent, given the amount of work involved in preparing them (as judged by the level of complaints in preparing and submitting them). For example:
  - *Schedule.* Here we find key schedule events and how well the program is doing against baseline thresholds and objectives, which can be used to help understand program timelines from Milestone B through Milestone C. This allows analysts to identify how long acquisition takes (cycle time) and any schedule growth.
  - *Performance.* This section lists baseline performance *requirements*, demonstrated performance to date, and estimated final performance. Analysis across SARs identifies whether requirements are changing (and if they are harder or easier to meet) and how well the acquired system is meeting the stated requirements of the warfighter.
  - *Track to budget.* This feature allows analysts to track where program funding is coming from across all types of appropriations (Research Development Test and Evaluation, Procurement, O&S, Military Construction, etc.). Without this mapping, it is hard to trace appropriations to programs across DoD because they are not aligned (e.g., budget "program elements" can fund multiple acquisition programs).
  - *Cost and funding.* Among the items in this section, the quantity-aligned unit costs allow analysts to build cost-quantity learning curves to understand how quickly (or even whether) costs for subsequent units are coming down in real (inflation-adjusted) terms. Also, these data are critical for adjusting unit procurement costs for quantity changes. Without adjusting for quantity, we cannot tell whether unit cost growth is due to requirement (quantity) changes or to poor acquisition system performance.
  - *Unit cost.* The unit costs are used to directly identify whether programs have breached congressionally mandated Nunn-McCurdy cost thresholds and the associated report-

ing, review, restructuring, or cancellation activities required by law.[6] The SAR record for a program also allows one to identify what baseline is used for a program's reported unit cost growth.

Options to Consider

We offer three options for DoD leadership to consider:

1. DoD should take this opportunity presented by Congress to reassess, improve, and streamline the current information contained in the SAR, the structure of the SAR itself, and the process by which this information is reported to Congress and DoD. The SAR itself does not necessarily need to be preserved, but DoD needs to continue to collect the program data it contains and disseminate the data to both internal and external stakeholders. DoD should also consider how the program managers and program executive officers can benefit from a new or revised version of this information requirement.
2. While exploring revisions or alternatives to the SAR, USD(A&S) should consider temporarily codifying in acquisition policy the requirement of submitting SARs to OSD while weighing changes or negotiating with Congress. Regardless of whether reports to Congress are reestablished, the preservation of the SARs will ensure that these data are available without a break to inform decisions and policymaking. Key data elements should retain their structures to enable continued trend analysis.
3. Concurrently, DoD should ask Congress to reconsider the legislation that terminates the submission of SARs until DoD has had a chance to consider how the information contained in the SARs and their submission can be improved.

---

[6] U.S. Code, Title 10, Section 2433, Unit Cost Reports.

CHAPTER FIVE

Whither the Defense Acquisition Executive Summary? An Opportunity to Align DAES to Support Emerging Acquisition Priorities

Recent changes to the organizations and RRAs for DoD acquisition have resulted in the DAES going partially dormant.[1] DAES has been a core mechanism enabling OSD and component insight and oversight of MDAPs. Historically, the DAES data and execution oversight process have been constructed and adjusted to meet the mission needs, goals, and preferences of the DAE.[2] Even with the delegation of most MDAP oversight to the components, the USD(A&S)—the DAE—has strong reasons to preserve, improve, and streamline some portions of DAES, especially the acquisition program data reporting, which support broader needs of policy and decisionmaking, governance, mission- and kill-chain portfolio analysis, and DEPSECDEF review across the DoD enterprise. Also, the DAES program execution oversight process could be transformed into a higher-level portfolio oversight function to inform decisionmaking and facilitate better coordination of program timing, budgeting, and strategic support to deliver warfighting capabilities.

Background and Motivation

The Strategy, Data, and Design (SDD) office was instituted by USD(A&S) with the mission of realigning offices within OUSD(A&S) to better support leadership's acquisition policy goals. In FY 2018, SDD asked the RAND National Defense Research Institute for guidance on whether and how to adapt DAES to the emerging acquisition environment. The paper on which this chapter is based responds to that request by synthesizing relevant findings and recommendations from a corpus of prior RAND work.[3]

Until recently, DAES information enabled insight into MDAP execution status by key component and OSD stakeholders between the major acquisition milestones. With congres-

---

[1] DAES comprises data, data collection, and program selection/presentation to the Defense Acquisition Executive (DAE). The data collection, program selection, and briefings to the DAE have traditionally been referred to as the "DAES process" by the workforce. We continue that nomenclature here, distinguishing between DAES data and process.

[2] The DAES process started under the Under Secretary of Defense (Acquisition). When the DAES process was initiated in the late 1980s, its primary purpose was to obtain early warning of potential program execution problems—largely technical or engineering-related—that could adversely affect cost, schedule, and performance. In 2005, new acquisition leadership changed the purpose of the DAES process to emphasize a more general program review focused on cost, schedule, and budget information. In 2010, leadership again changed the primary purpose to a hybrid model, with both program execution review and risk identification.

[3] For additional background on the DAES process, along with challenges and options for improving the process, see Drezner et al., 2018. See also Riposo et al., 2015; McKernan et al., 2016; McKernan et al., 2017; Drezner et al., 2019; and McKernan et al., 2018.

sional delegation of MDA and an evolving requirement for strategic insight by A&S leadership, a portion of the DAES process has been discontinued. The data associated with the DAES report itself are still collected, but the monthly DAES reviews no longer occur.

DAES employs a data structure similar to the SAR but is supplemented with additional insight through program assessments from both program managers and functional leads in multiple OSD offices. Along with the SAR, the DAES data have been one of the only sources of consistent, longitudinal, standardized program information supporting program, portfolio, process, and policy analysis for the almost $100 billion in annual investment in MDAPs.[4] The structure and data definitions associated with DAES also constitute an important starting place for the development of common acquisition program data management across all program levels, program types, and components (i.e., potentially part of common enterprise acquisition data).[5]

Section 825(c)(2) of the NDAA for FY 2016 provides that "the [MDA] for [an MDAP] reaching Milestone A after October 1, 2016, shall be the service acquisition executive [SAE] of the military department that is managing the program, unless the Secretary of Defense designates . . . another official to serve as the [MDA]."[6] In addition, MDA has been delegated to the SAEs for most existing MDAPs.[7] Currently, the USD(A&S) is the MDA for only 11 ACAT ID programs. The SAEs are MDAs for the remaining 87 MDAPs.[8] This delegation of new and preexisting MDAPs to the military departments has left the DAES without an execution-oversight purpose for most MDAPs. The delegation of MDA also came with an explicit direction to minimize documentation reporting outside the components.[9]

---

[4] Office of the Under Secretary of Defense(Comptroller)/Chief Financial Officer, 2018.

[5] U.S. Code, Title 10, Section 2222, Defense Business Systems: Business Process Reengineering; Enterprise Architecture; Management.

[6] Pub. L. 114-92, 2015; U.S. Code, Title 10, Section 2430(d), Major Defense Acquisition Program Defined.

[7] Under Secretary of Defense for Acquisition, Technology, and Logistics, "Army Program Delegation Decisions Acquisition Decision Memorandum," memorandum, Washington, D.C.: U.S. Department of Defense, March 17, 2017a, Not available to the general public; Under Secretary of Defense for Acquisition, Technology, and Logistics, "Navy Program Delegation Decisions Acquisition Decision Memorandum," memorandum, Washington, D.C.: U.S. Department of Defense, March 20, 2017b, Not available to the general public; Under Secretary of Defense for Acquisition, Technology, and Logistics, "Air Force Program Delegation Request Acquisition Decision Memorandum," memorandum, Washington, D.C.: U.S. Department of Defense, June 30, 2017c, Not available to the general public.

[8] The number of ACAT I (IB, IC, and ID) programs that are "actively reporting" is 98 as of October 22, 2018. This information is presented in the authoritative list of acquisition programs in the DAVE.

[9] 10 U.S.C. 2430(d)(3)(B) says that

> The Secretary of Defense shall review the acquisition oversight process for major defense acquisition programs and shall limit outside requirements for documentation to an absolute minimum on those programs where the service acquisition executive of the military department that is managing the program is the milestone decision authority and ensure that any policies, procedures, and activities related to oversight efforts conducted outside of the military departments with regard to major defense acquisition programs shall be implemented in a manner that does not unnecessarily increase program costs or impede program schedules.

Consequences of a Partially Dormant Defense Acquisition Executive Summary

A partially dormant DAES has two fundamental consequences for acquisition data management:[10]

1. The components may halt data reporting altogether. This action would eliminate a critical authoritative source of program information, adversely affecting longitudinal and portfolio analyses, policymaking, and understanding of the performance of the overall acquisition system.
2. The common data framework underlying DAES reporting would be weakened, potentially leading to a divergence and inconsistency in how and what program data are tracked, reported, and used for decisionmaking across DoD. This divergence would inhibit the ability of the DAE and SAEs to maintain consistent information on program execution status, portfolio status, effectiveness of policy, and the overall performance of the acquisition system to the SECDEF, the DEPSECDEF, and Congress.

The data generated by the legacy DAES process include standardized cost, schedule, performance, contracting, and budget data on all active MDAPs, along with more-detailed assessments explaining changes in 11 functional areas.[11] These data were supplemented by program risk and issue summary charts, as well as a systems engineering reliability growth curve. Additional information generated during the process included briefings by the Overarching Integrated Product Team lead, Program Executive Officer, and PM for individual MDAPs.

Elimination of this information source will, in turn, eliminate the benefits that have accrued from its use. Of particular concern is the potential loss of consistent data for measuring system, institutional, and program performance (as is also the case with the termination of the SAR to Congress).[12] Without these standards, the components' performance measurements may drift over time, leading to significant confusion and inefficiencies, a reduced ability to integrate business practices across DoD, and reduced transparency. In addition, the DAES has been a forcing function for continuous communication between the components and OSD. These communication lines will likely atrophy, along with timely OSD knowledge of the performance of both individual programs and the overall defense acquisition system. However, also at issue is whether the Services should continue to submit information to DAES and whether a common reporting requirement is still needed. From an analyst's viewpoint, such data are a treasure trove and should be preserved. From the Service perspective, DAES may be burdensome and may not be helpful to the internal management or oversight of the program within the military service itself. The question of whether broad use of such data across the entire defense acquisition establishment is useful and pertinent needs to be considered by OSD and Service leadership.

The elimination of the position of Under Secretary of Defense for Acquisition, Technology and Logistics and the creation of two new principal positions—USD(R&E) and USD(A&S)—

[10] Drezner et al., 2018.

[11] These data can be found in the DAMIR information system.

[12] Jeffrey A. Drezner, Philip S. Antón, Megan McKernan, and Jerry M. Sollinger, unpublished RAND Corporation research, 2018.

is a complicating factor that will also affect DAES data. Key stakeholders who provide data inputs (e.g., functional-area assessments in OSD) now report to different organizations; for example, the Deputy Assistant Secretary of Defense (DASD) for Systems Engineering and the DASD for Developmental Test and Evaluation report to USD(R&E), not USD(A&S). Those officials may prioritize or formulate data needs differently, potentially creating a misalignment between R&E and A&S data requirements and data management. Continued management of acquisition data through the existing DAES data framework has the potential to maintain data alignment between A&S and R&E. Incremental changes to the data framework could address new or emerging organizational missions and priorities.

The duties of the USD(A&S), as codified in Section 133b of U.S.C. Title 10, require the type of information collected and maintained in DAES to inform policymaking and to facilitate understanding of portfolio issues. In particular, the USD(A&S)'s role of chief A&S officer—being the principal adviser to the SECDEF on acquisition—involves establishing policies that enable an efficient and effective acquisition system. In principle, policymaking requires the kind of information contained in the DAES. USD(A&S) also has an advisory role when the SAE is the MDA and requires some data to inform that advice.

Opportunities to Improve Acquisition Data Collection and Add Strategic Value

Some DAES data elements supply critical information for both USD(A&S) and the components and could be improved or streamlined; less-critical data could be eliminated. Below are some example opportunities for further analysis and consideration by both OSD and SAE-level leadership.

- **Cost, schedule, and performance.** These DAES sections provide traditional metrics for cost, schedule, and performance that are identical in structure to those in the SAR and provide both baselines and current estimates. They constitute the core information used for program status at various ACAT levels and across the components and OSD. They are also updated quarterly, providing leadership and analysts with more-timely information than the annual SARs (which currently are repealed as of December 2021).[13]
- **Program manager and OSD assessments.** These DAES assessments provide a unique data set of short statements of status and issues in 11 functional areas: program cost, program schedule, system performance, funding, test and evaluation, sustainment, management, contract performance, interoperability/information security, production, and international program aspects. This information captures both the component- and OSD-level views of current program status, drawing out potential differences. For analysts, this information helps reduce data calls to the program offices for current status. Merging topics may ease the burden on both component and OSD staff, which have been reduced by mandatory cuts.
- **Risk and issue summaries.** As with the PM and OSD assessments, the DAES risk and issue summaries offer a unique dataset. The risk and issue summaries provide current major risks/issues and any plans/schedule to mitigate them, but they are separate files

[13] Pub. L. 115-91, Sec. 1051(x)(4), 2017.

somewhat hidden in the DAMIR system and, thus, are less known and less used than they could be. These datasets may be worth preserving, given the importance of identifying risks in acquisition, and they could be integrated into the traditional DAES report.
- **Contracting, production, and earned value.** DAES reporting currently includes metrics on both overall program and contractor performance using various measures, such as earned value, that are not reported in the SAR. Top-level earned-value information provides leadership and analysts with additional information from the prime contractor in multiple metrics to understand whether work is being completed on time and within budget. Earned-value data are automatically extracted from the Earned Value Management Central Repository, so their inclusion provides useful insight without a reporting burden on the programs. Also included is some additional contract performance information from multiple assessments completed in part by the Defense Contract Management Agency (DCMA), whose analysts have firsthand knowledge of potential production and quality assurance issues with the prime contractors.

Given the refocus of USD(A&S)'s mission, additional strategic value can be created by using information from DAES for portfolio analysis and reviews. With the expanded emphasis in DoD on understanding how programs integrate in portfolios to produce operational capabilities and effects, DAES could be expanded and structured to present various views and insights for different types of portfolios (e.g., by kill chain; by program interdependencies; by key commodity classes, such as anti-access area denial). Insights could include timing inconsistencies (e.g., programs that will not deliver in time to support other weapon systems) and budgetary and quantity inconsistencies (e.g., cuts to a program that affect other weapon systems that together provide an important operational capability). However, these issues need higher-level resolution, either by USD(A&S) or the SAEs. The data discussed are pertinent and valuable to an oversight role, and if the Services are already tracking by portfolio, then it would be better to share those data than to institute an additional data requirement.

Recommended Actions

Since DAES exists for USD(A&S) and her staff, we recommend that the opportunity be taken to rethink DAES—e.g., the data that are collected, how they are used, and the structure and operation of the process—to support the mission needs, goals, and policy preferences of the DAE and the SAEs.

- Regardless of how the DAES process oversight and insight evolves, we recommend that the data reporting structure under DAES be preserved, maintaining that data framework in concert with the SAEs. We also recommend building on the foundation the legacy DAES data provide; the data still being reported are very important for decisionmaking, policymaking, governance, and mission- and kill-chain portfolio analysis beyond execution oversight. The structure and data definitions associated with the DAES constitute an important starting place for the development of common acquisition program data management across all program levels, program types, acquisition processes, and components.
- Reengineering DAES should take into account the many interdependencies associated with multiple data uses, including congressional reporting, component and OSD over-

sight and insight functions, longitudinal analyses, and mission- and kill-chain portfolio analysis. Alignment can also be created between A&S and R&E for common uses.

- Data collection should be evaluated in light of supporting USD(A&S)'s strategic goals of policymaking, governance, and mission- and kill-chain portfolio analysis. This evaluation could include an examination of other existing information flows (e.g., the Air Force's Monthly Acquisition Reports; DCMA's quarterly Program Assessment Reports; Earned Value Management data) and whether they can be integrated or automated to provide a DoD-wide execution status and issue-raising forum. This holistic, enterprise-wide assessment of program data and other related acquisition information requires USD(A&S) direction, leadership, and support, to provide a strong foundation for designing effective acquisition policy going forward.

CHAPTER SIX

Conclusions

Acquisition program data and the analytics they support are critical to acquisition decisionmaking. The reorganization of OSD acquisition-related functions and changes in RRAs created challenges for data governance and management. Senior leadership's attention can turn these challenges into opportunities to rethink data requirements and lay a strong foundation for data-driven decisionmaking into the future.

The short papers presented in this report address only a few of the challenges associated with acquisition program data governance and management due to the OSD reorganization and changes in RRAs. Even without being exhaustive, the papers identify several fundamental data-related issues requiring senior leadership attention. Viewed another way, these changes in organizational structure and RRAs create opportunities to rethink acquisition program data requirements to better support acquisition decisionmaking in both the Services and OSD at the program, integrated-capability, policy, and strategic levels.

Key Observations

As with any large, complex organization, DoD faces challenges related to data access and management. Prior to the current reorganization and statutory changes, the challenges affecting acquisition information included complex security policies regulating information systems; cultural and technical barriers to accessing and sharing information; and lack of awareness of the breadth and depth of information available to DoD leaders and staff. A rich set of information is available to support acquisition insight, analysis, and decisionmaking, but the full extent to which this information is used remains unknown. In addition, no common data environment exists for all acquisition information, and there is little agreement on all data needs and definitions across DoD. Both issues result from decentralized governance and management. While most information used for program management and oversight/insight is similar across OSD and the Services (at least for similarly sized programs), specific performance metrics and uses differ. Recent statutory changes to organizational structures and RRAs will exacerbate these challenges and have introduced new challenges and opportunities that will significantly affect the collection, storage, and use of acquisition information.

The new Middle Tier acquisition pathway will require program data to inform both programmatic and policy decisions. Middle Tier acquisition will need to address and resolve many of the challenges that have faced traditional acquisition processes in the past. These challenges include the following:

- Exactly what program data are reported, at what frequency, and how must be determined. While the A&S and Service guidance memos address this issue, they do not resolve it.
- Data reporting is uncoordinated across organizations.
- The Service guidance memos reflect a lack of standardization across organizations in terms of what should be reported, relying instead on tailoring data reporting to reflect the characteristics of each program. No guidance is provided on how to tailor or how to determine what is appropriate for a given program.
- The objective of the Middle Tier pathway is speed. Reporting requirements risk overburdening the process and slowing it down.

The Middle Tier acquisition pathway also illustrates some of these data opportunities. In particular, it demonstrates how the existing data infrastructure (information systems, data-collection conventions, common data definitions) can support and adapt to new acquisition authorities and processes.

The SAR and DAES represent a common data framework for ACAT I programs, and recent initiatives in the Services suggest that the framework is being applied to the smaller ACAT II–IV programs.[1] Elimination of this information source by Congress will, in turn, eliminate many of the benefits that have accrued from its use over time. Of particular concern is the potential loss of common data standards and definitions for measuring program performance. Without these common data standards and definitions (i.e., a common data framework), institutionalized over decades of SAR creation and submission, the military departments' performance measurements may drift over time, leading to significant confusion and inefficiencies, a reduced ability to integrate business practices across DoD, and reduced transparency.

Key stakeholders who provide data inputs (e.g., functional-area assessments in OSD) now report to different organizations; for example, the DASD (Systems Engineering) and the DASD (Developmental Test and Evaluation) report to USD(R&E), not USD(A&S). Those officials may prioritize or formulate data needs differently, potentially creating a misalignment between R&E and A&S data requirements and data management. Continued management of acquisition data through the existing DAES data framework has the potential to maintain data alignment between A&S and R&E. Incremental changes to the data framework could address new or emerging organizational missions and priorities.

Options to Consider Going Forward

To mitigate any unintended consequences from these recent statutory changes, USD(A&S) and the SAEs could address these challenges as policies are updated to reflect the emerging acquisition environment, as well as issue new policy and guidance to address new challenges and opportunities.

- **USD(A&S) and the SAEs could create a strategic management plan for acquisition information that determines what information is needed by whom to accom-**

[1] Data elements for both the SAR and DAES can be found in OUSD(A&S)'s Acquisition Visibility Data Matrix, located in DAVE.

plish enterprise-wide objectives without overburdening the military departments. This near-term guidance could help keep the required data flows from atrophying while also providing the foundation for rethinking data needs to better reflect emerging policy emphasis on acquisition process governance, program insight, and mission- and kill-chain portfolio analysis.
- **USD(A&S) and the SAEs could work together to standardize a core set of data definitions, authoritative sources, and management approaches.** Standardized data definitions would facilitate information sharing and understanding; align data governance and management across organizations, uses, and program types; and be an important substantive step toward a common acquisition data framework.

One opportunity to address these challenges will be the collaborative policymaking process directed in the USD(A&S) interim guidance memorandum. Using prior research, we offer the following four guidelines to ensure that requirements and processes associated with Middle Tier program data and other acquisition information are as efficient and effective as possible:

1. **Let decisionmaking drive data requirements.** Data and information must not be generated for their own sake but must support important decisionmaking about policy, process, programs, and integrated capability outcomes.
2. **Minimize reporting requirements.** Information and documentation requirements should be austere, with minimal data reporting. Historically, successful rapid prototyping and fielding activities have had austere information requirements. Guidance appears to recognize this by emphasizing tailoring.
3. **Standardize where possible.** A common acquisition program data framework should be developed for a core set of program data. The existing data framework reflected in the legacy SAR and DAES provides a strong foundation from which to start.
4. **Capitalize on existing structures.** Existing data frameworks, information systems, and organizations should be used to the maximum extent practical, especially when such data are shared automatically between systems.

Since the DAES process is designed to support the DAE, now the USD(A&S), we recommend taking the opportunity to rethink DAES—the data that are collected, how they are used, and the structure and operation of the process—to support the mission needs, goals, and policy preferences of the DAE. Regardless of how DAES process oversight and insight evolves, we recommend preserving the data reporting structure under DAES and maintaining that data framework in concert with the SAEs. The structure and data definitions associated with DAES constitute an important starting place for the development of common acquisition program data governance and management across all program levels, program types, acquisition processes, and Components. Reengineering of DAES should encompass the many interdependencies associated with multiple data uses, including congressional reporting, Component and OSD oversight and insight functions, longitudinal analyses, and mission- and kill-chain portfolio analysis. Alignment can also be created between A&S and R&E for common uses of data.

Similarly, USD(A&S) should take the opportunity presented by Congress to reassess, improve, and streamline the information contained in the SAR, the structure of the SAR itself, and the process by which this information is reported to Congress and DoD. Like the legacy DAES, the SAR itself does not necessarily need to be preserved, but the program data it contains need to continue to be collected and disseminated to both DoD and external stakeholders. Historically, the data reported in the legacy SAR and DAES have been largely aligned; this historical alignment suggests that reengineering these documents and the data they contain should be done in parallel. The core data requirements for a range of uses—from congressional reporting to mission- and kill-chain portfolio analysis—are likely to be the same, offering some degree of savings and enhanced data-governance and data-management practices.

Bibliography

Assistant Secretary of the Air Force for Acquisition, Technology and Logistics, "Seven Steps for Incorporating Rapid Prototyping into Acquisition," memorandum, Washington, D.C.: Department of the Air Force, April 10, 2018a.

———, "Air Force Guidance Memorandum for Rapid Acquisition Activities," memorandum, Washington, D.C.: Department of the Air Force, June 13, 2018b.

Assistant Secretary of the Army for Acquisition, Logistics and Technology, "Office of the Assistant Secretary (Acquisition, Logistics and Technology) of the Army Middle-Tier Acquisition Policy," memorandum, Washington, D.C.: Department of the Army, September 25, 2018. As of April 25, 2019: https://aida.mitre.org/wp-content/uploads/2018/10/MTA-ASAALT-Policy-25-Sep-18-3.pdf

Assistant Secretary of the Navy for Research, Development, and Acquisition, "Middle Tier Acquisition and Acquisition Agility Guidance," memorandum, Washington, D.C.: Department of the Navy, April 24, 2018.

Bolten, Joseph G., Robert S. Leonard, Mark V. Arena, Obaid Younossi, and Jerry M. Sollinger, *Sources of Weapon System Cost Growth: Analysis of 35 Major Defense Acquisition Programs*, Santa Monica, Calif.: RAND Corporation, MG-670-AF, 2008. As of October 17, 2018: https://www.rand.org/pubs/monographs/MG670.html

Comptroller General of the United States, "Response to the Honorable F. Edward Herbert, Chairman, Committee on Armed Services, House of Representatives," B 163058, October 30, 1973.

DoD—*See* U.S. Department of Defense.

Drezner, Jeffrey A., and Meilinda Huang, *On Prototyping: Lessons from RAND Research*, Santa Monica, Calif.: RAND Corporation, OP-267-OSD, 2009. As of May 25, 2018: https://www.rand.org/pubs/occasional_papers/OP267.html

Drezner, Jeffrey A., Megan McKernan, Badreddine Ahtchi, Austin Lewis, and Douglas Shontz, *Issues with Access to Acquisition Data and Information in the Department of Defense: Streamlining and Improving the Defense Acquisition Executive Summary (DAES) Process and Data*, Santa Monica, Calif.: RAND Corporation, 2018, Not available to the general public.

Drezner, Jeffrey A., Megan McKernan, Austin Lewis, Ken Munson, Devon Hill, Jaime Hastings, Geoffrey McGovern, Marek Posard, and Jerry M. Sollinger, *Issues with Access to Acquisition Data and Information in the Department of Defense: Identification and Characterization of Data for Acquisition Category (ACAT) II–IV, Pre-MDAPs, and Defense Business Systems*, Santa Monica, Calif.: RAND Corporation, 2019, Not available to the general public.

Lord, Ellen, "Department of Defense Press Briefing on DoD Acquisition Reforms and Major Programs," Office of the Under Secretary of Defense (Acquisition and Sustainment), May 10, 2019.

McKernan, Megan, Jeffrey A. Drezner, and Jerry M. Sollinger, *Tailoring the Acquisition Process in the U.S. Department of Defense*, Santa Monica, Calif.: RAND Corporation, RR-966-OSD, 2015. As of July 25, 2019: https://www.rand.org/pubs/research_reports/RR966.html

McKernan, Megan, Nancy Young Moore, Kathryn Connor, Mary E. Chenoweth, Jeffrey A. Drezner, James Dryden, Clifford A. Grammich, Judith D. Mele, Walter T. Nelson, Rebeca Orrie, Douglas Shontz, and Anita Szafran, *Issues with Access to Acquisition Data and Information in the Department of Defense: Doing Data Right in Weapon System Acquisition*, Santa Monica, Calif.: RAND Corporation, RR-1534-OSD, 2017. As of July 25, 2019:
https://www.rand.org/pubs/research_reports/RR1534.html

McKernan, Megan, Jessie Riposo, Jeffrey A. Drezner, Geoffrey McGovern, Douglas Shontz, and Clifford A. Grammich, *Issues with Access to Acquisition Data and Information in the Department of Defense: A Closer Look at the Origins and Implementation of Controlled Unclassified Information Labels and Security Policy*, Santa Monica, Calif.: RAND Corporation, RR-1476-OSD, 2016. As of July 25, 2019:
https://www.rand.org/pubs/research_reports/RR1476.html

McKernan, Megan, Jessie Riposo, Geoffrey McGovern, Douglas Shontz, and Badreddine Ahtchi, *Issues with Access to Acquisition Data and Information in the Department of Defense: Considerations for Implementing the Controlled Unclassified Information Reform Program*, Santa Monica, Calif.: RAND Corporation, RR-2221-OSD, 2018. As of July 25, 2019:
https://www.rand.org/pubs/research_reports/RR2221.html

Office of the Under Secretary of Defense (Comptroller)/Chief Financial Officer, *Program Acquisition Cost by Weapon System, Fiscal Year 2019 Budget Request*, Washington, D.C.: U.S. Department of Defense, February 2018.

Office of the Under Secretary of Defense for Acquisition and Sustainment, "Background of SAR," Washington, D.C.: U.S. Department of Defense, accessed via the Defense Acquisition Management Information Retrieval, as of October 3, 2018.

OUSD(A&S)—*See* Office of the Under Secretary of Defense for Acquisition and Sustainment.

Public Law 114-92, National Defense Authorization Act for Fiscal Year 2016, November 25, 2015.

Public Law 114-328, National Defense Authorization Act for Fiscal Year 2017, December 23, 2016.

Public Law 115-91, National Defense Authorization Act for Fiscal Year 2018, December 12, 2017.

Riposo, Jessie, Megan McKernan, Jeffrey A. Drezner, Geoffrey McGovern, Daniel Tremblay, Jason Kumar, and Jerry M. Sollinger, *Issues with Access to Acquisition Data and Information in the Department of Defense: Policy and Practice*, Santa Monica, Calif.: RAND Corporation, RR-880-OSD, 2015. As of July 25, 2019:
https://www.rand.org/pubs/research_reports/RR880.html

Under Secretary of Defense for Acquisition and Sustainment, "Middle Tier of Acquisition (Rapid Prototyping/Rapid Fielding) Interim Authority and Guidance," memorandum, Washington, D.C.: U.S. Department of Defense, April 16, 2018a.

Under Secretary of Defense for Acquisition and Sustainment, "Middle Tier of Acquisition (Rapid Prototyping/Rapid Fielding) Interim Guidance," memorandum, Washington, D.C.: U.S. Department of Defense, October 9, 2018b.

Under Secretary of Defense for Acquisition and Sustainment, "Middle Tier of Acquisition (Rapid Prototyping/Rapid Fielding) Interim Guidance 2," memorandum, Washington, D.C.: U.S. Department of Defense, March 20, 2019.

Under Secretary of Defense for Acquisition, Technology, and Logistics, "Army Program Delegation Decisions Acquisition Decision Memorandum," memorandum, Washington, D.C.: U.S. Department of Defense, March 17, 2017a, Not available to the general public.

Under Secretary of Defense for Acquisition, Technology, and Logistics, "Navy Program Delegation Decisions Acquisition Decision Memorandum," memorandum, Washington, D.C.: U.S. Department of Defense, March 20, 2017b, Not available to the general public.

Under Secretary of Defense for Acquisition, Technology, and Logistics, "Air Force Program Delegation Request Acquisition Decision Memorandum," memorandum, Washington, D.C.: U.S. Department of Defense, June 30, 2017c, Not available to the general public.

U.S. Code, Title 10, Armed Forces, release point 115-137, March 16, 2018.

U.S. Code, Title 10, Section 2222, Defense Business Systems: Business Process Reengineering; Enterprise Architecture; Management.

U.S. Code, Title 10, Section 2430, Major Defense Acquisition Program Defined.

U.S. Code, Title 10, Section 2433, Unit Cost Reports.

U.S. Code, Title 10, Section 2440, Technology and Industrial Base Plans.

U.S. Code, Title 10, Section 2447, Weapon System Component or Technology.

USD(A&S)—*See* Under Secretary of Defense for Acquisition and Sustainment.

U.S. Department of Defense, *Report to Congress Restructuring the Department of Defense Acquisition, Technology and Logistics Organization and Chief Management Officer Organization, in Response to Section 901 of the National Defense Authorization Act for Fiscal Year 2017 (Public Law 114-328)*, Washington, D.C., August 2017.